2

3

4

5

6

7

papua new guinea

BIRDS

In colour, Papua New Guinea birds rival or surpass those of other countries, with the birds of paradise often stated to be the most beautiful birds in the world.

Papua New Guinea is only one tenth the size of Australia and only one twenty-fifth the size of North America but has a comparable number of species of birds - 740. Nearly one fifth of these are migrants. Seventy-seven species are found in Papua New Guinea and nowhere else. They have evolved here through selection of characters best suited to the local environment which itself has evolved since the whole island of New Guinea, and its nearby satellite islands, rose from the sea less than 15 million years ago.

The birds of PNG are most closely related to those of Australia. In fact, in the open savannah woodlands of the south coastal regions, about ninety per cent of the bird species are Australian.

For an understanding of the diverse variety of the birds of PNG it is necessary to understand the diverse nature of the environment. Within the island of New Guinea we have habitats ranging from coral atolls to humid tropical rainforests, dry woodland savannahs, cold upland mossy forests, high mountain grasslands and even snowcapped mountain tops.

The vegetation of PNG is divided into five or six altitudinal zones and to a large extent the birds are divided in the same way. For instance there are three species of colourful ground birds called Jewel-babblers. The Blue Jewel-babbler is found in the lowland forests up to 800m. The Chestnut-backed Jewel-babbler is found in the hill forests and lower montane forests from 900m to 1400m and the Spotted Jewel-babbler is found in the mid to upper montane forests from 1500m to 1500m. This closely matches the division of the vegetation zones.

This booklet is intended to stimulate the reader's curiosity to find out more about PNG birds and to give a lead on how to do this. Even a householder in one of PNG's towns such as Port Moresby, Lae or Mt Hagen will have many opportunities to observe birds perching or feeding in the trees or gardens or maybe just flying overhead. Even in the towns new information can be found about PNG's birds. For example, Hagenites will have observed the flocks of smallish black birds which congregate on the radio tower above the post office. They are Singing Starlings *Aplonis cantoroides*, and the bird books say that this bird is a lowland species which may occasionally get as high as 1500m. Mt Hagen is almost 1700m so that is new information to be added to our knowledge of this bird. If we know the bird's habits we can understand why this bird is in Mt Hagen. It is a bird of open country and will take up residence in newly-opened areas. It also uses man's environment for roosting and nesting. This bird is a native species and should not be confused with the European Common Starling.

Not all areas are good for birds but even the worst kind of habitat will have some birds. New birdwatchers want to see lots of birds so they need to go to areas where it is known by other 'birders' that there are plenty to see. First it is a good idea to be familiar with your own home area birds. In Port Moresby there is an organisation which can help you considerably; the PNG Bird Society. The members of this society make trips to good 'birding' areas at least once a month as a group. This is a good way to learn the birds quickly. In other towns it is harder unles you can team up with other like-interested persons.

However there are several well-known areas where visiting tourist/birdwatchers are taken to see some of our outstanding and colourful birds. For instance there is a wide variety of habitats around Port Moresby. If you go boating out to Fisherman's (Daugo) Island or to Haidana Island you can see several seabirds and shorebirds (waders). If you go to Varirata National Park you can see many forest birds including the Raggiana Bird of Paradise (ask one of the Rangers). If you go to the Waigani Swamps or inland from Lea Lea to Lake Iraguma you can see a wide range of swamp and lake birds including egrets, pelicans, pied herons, terns, waterhens and many ducks. Even in town you can see honeyeaters, butcherbirds, friar-birds, lorikeets and flycatchers.

If you wish to see some of the highland birds you should travel to the Tari Gap or to Tomba on the Hagen-Wabag road. Several birds of paradise are readily seen there especially the Ribbontail, Brown Sicklebill, King of Saxony varieties and the Short-tailed Paradigalla.

Along the Sepik River and Karawari River and their adjacent lakes on the north side of PNG there are masses of waterbirds to be seen. On the south side of PNG probably the best area for water and swamp birds is the Bensbach Plains in the Tonda Wildlife Management area. It is difficult to get to but most rewarding, especially from April to September.

There are three species of cassowary in PNG as shown here, one of which, the double-wattled Cassowary *Casuarius casuarius* is also found in Queensland, north Australia, where it is under threat of extinction because of the destruction of the remaining stands of rainforest. In general terms the distribution of the cassowaries is that the Double-wattled is found in the southern lowlands and hill forests up to 500m and in the northeastern provinces. The Dwarf or Bennett's Cassowary *Casuarius bennetti* is found mostly in the mountains of PNG but is found in the lowlands of the southeast. The Single-wattled Cassowary *Casuarius unappendiculatus* is found in the lowland forest of the northwest from the Ramu River valley west into West Irian. It is not unusual to find either of these species outside their normal range for there is a large trade in young cassowaries around the country. Cassowaries belong to the same family of birds as the Australian emu, the African ostrich and the South American rhea - the ratites. The cassowaries are well known as fruit-eaters, picking up the fruits fallen to the ground from a large number of tree species. They also eat a lot of meat in the way of frogs, lizards, rats and the young of ground nesting birds. (1, 2 & 3)

1

Pelicans *Pelecanus conspicillatus* are very common in the far southwestern corner of PNG along the Bensbach River and its associated vast swamps and lagoons. However, sometimes large numbers of pelicans come up from Australia and spread out all over PNG and its island archipelagoes even up to the lakes and rivers in the highlands in their search for fish. They are frequently seen on the lakes to the north of Port Moresby. (4)

Among the most graceful birds in the world are the herons and egrets; so often the subject of artists. PNG has a good variety of species including the White-faced Heron *Notophoyx novaehollandiae* an uncommon but widely-distributed bird here. It is more common in Australia. Most of PNG's waterbirds are Australian birds. (5)

One of the rarer herons in PNG is the Pacific or White-necked Heron *Ardea pacifica* which like the other herons inhabits the swamplands and lakes. It is a vagrant from Australia but may be found in the highlands as much as in the lowlands. It is one of the larger herons and easily identified by the row of dark feathers down the front of the neck. (6)

Except for the egrets the Pied Heron *ardea picata* is probably the commonest species of this group to be seen in PNG. It is found in most lowland swamps and lakes and along rivers and mostly occurs in flocks of ten to a hundred or more. Young birds do not have the black cap but are white all over the head and neck. (7)

The Great Egret *Egretta alba* is almost cosmopolitan and quite a common bird in PNG particularly in the lowlands. It is the tallest of the five white egret species in PNG but is not always easy to identify when it is mixed up with hundreds of other egrets. However, it seldom folds its neck down and the feathers of the neck are short and so it does not get the hunched appearance of the other species. (8)

A most maligned bird is the stork. There is only one species in PNG, the Black-necked Stork or Jabiru *Xenorhynchus asiaticus*. It may be seen commonly only in the southwest in the Trans-Fly and Bensbach areas. It is also found from southern Asia, Indonesia and Australia. (9)

The Brolga or Native Companion *Grus rubicunda* is the only species of the crane family in PNG. It only occurs in the Trans-Fly/Bensbach area where it is very common. It is also common in Australia where another closely-related species the Sarus Crane is also found. It could be found here in PNG and could be identified by the red colour of the face of the Brolga extending part way down the neck of the Sarus. (10)

The commonest duck by far in PNG is the Pacific Black Duck *Anas superciliosa*. It is found all over the mainland up to the high mountain lakes and to the surrounding islands. It is also found in Indonesia, Australia, New Caledonia and as far south as Macquarie Island. (11)

The Spotted Whistling Duck *Dendrocygna guttata* is common in lowland swamps and lakes but is not found on the north coast. It is one of three species of Whistling Ducks found in PNG and is not found in Australia. It can be recognised from the other two by the lack of long flank plumes. (12)

Note the long flank plumes on the Wandering Whistling Ducks *Dendrocygna arcuata*. It is common in lowland swamps and lakes on both sides of the mainland and is usually seen in large flocks either feeding or standing on a lake edge. It also occurs from Malaysia and Indonesia to Australia. (13)

The Pied or Magpie Goose *Anseranas semipalmata* is a very interesting bird. In PNG it is found only in the far southwest in the Bensbach area but is more common in the northern regions of Australia. It spends most of its time in swamps, lagoons and lakes and in wet grasslands. (14)

9

10

13

1

12

14

15

17

16

18

A bird of the wet lowlands and hill forests and regrowth areas, the red-necked Rail *Rallina tricolor* is seldom seen in daytime. It usually rests during the day but may be seen just on dusk as it moves out to feed in the forest or river banks. There are several closely similar species so check your field guide. (15)

There are several birds, known as rails, in the wet forest and along river banks of which the commonest is the Banded Landrail *Rallus philippensis*. It is most easily seen while driving along bush roads when it dashes off into the roadside forest or tall grasses. It is a secretive bird as are all the rails and occurs from the Philippines to Indonesia, all of New Guinea, Australia and some Pacific islands. (16)

The Spurwing or Masked Lapwing *Vanellus miles* is found in most southern lowland areas west of Port Moresby and occasionally in the north. It occurs near swamps, salt marshes, lakes and short grass fields such as outside Parliament House. Sometimes the Australian race with black nape and black streak on each side of the breast can be seen near Port Moresby. The one shown here is an intermediate form. (17)

Of the four species of mound-building megapodes the Wattled Brush-Turkey *Aepypodius arfakianus* is one of the commonest especially in the higher altitudes. Its nest mound is much smaller than that of the Common Scrubfowl found in coastal areas. It is only the male that has the red wattle hanging on its neck. (18)

This is one of the lesser-known eagles of the world, Gurney's Eagle *Aquila gurneyi* and it is only found on the island of New Guinea and its satellite islands. It occurs mostly near the coast but the one shown here was found near Banz in the Western Highlands at 1700m. It is nowhere common. (19)

The Wedge-tailed Eagle *Aquila audax* is one of the country's largest eagles but is only found in the far southwest of PNG. It is very common in Australia. Also in the southwest is the White-breasted Sea Eagle which, when young, is very similar to the Wedge-tail. (20)

This is one of PNG's smallest hawks, the Australian Kestrel *Falco cenchroides* which is sometimes seen commonly all over the country; other times rarely seen, occasionally absent. So far only females have been seen in PNG, which means there are no breeding records for this species in PNG. It feeds on insects and mice and will sometimes hover over grasslands waiting for prey to expose itself. (21)

The Brown Falcon *Falco berigora* is one of the commonest of our hawks. It is very vocal on the wing uttering short barks like a dog. It feeds mainly on rodents, lizards, snakes and insects and feeds on the ground. I once saw three of these birds hopping along the ground snatching at grasshoppers which were hatching from the ground in new short grass. (22)

20

21

22

23

25

24

26

The Bar-shouldered Dove *Geopelia humeralis* is a very common bird close to the coast from Port Moresby to the Trans-Fly area. It is often associated with mangrove forests from where it comes out to feed among the weeds and short grass nearby. It is often to be seen in suburban gardens and villages near the coast. (23)

PNG has a large variety of spectacularly-coloured pigeons, especially the group known as fruit doves. Here the Superb Fruit Dove *Ptilinopus superbus* is one of the most attractive. It is only 22cms long and it is amazing just how many different colours can be brought into harmony in such a small bird. It is common all over the lower altitudes of PNG up to 1800m. It sometimes nests only 2m from the ground. (24)

The bird shown here is a coastal bird found from the Nicobar Islands near India through Indonesia and New Guinea to the Solomon Islands. The Nicobar Pigeon *Caloenus nicobarica* is one of those coastal birds which feed on the mainland during the day and flies to offshore islands to roost for the night. (25)

Bird migration is a fascinating study and one good example of a migrating bird is the Torres Strait or Pied Imperial Pigeon *Ducula bicolor*. It is found in the coastal forests particularly in woodland savannahs where it feeds during the day. In the evenings it flies to nearby coastal islands to roost for the night. In early summer (November) large numbers migrate west along the south coast of PNG and fly across the Torres Strait to Australia, returning to PNG in March or April. Not all the birds migrate; there is a resident population. (26)

The Vulturine or Pesquet's Parrot *Psittrichas fulgidus* is our only species of protected parrot. It is becoming rare as its black and red plumage is much prized for head decoration at ceremonial *singsings*. The short stiff feathers of the head allow the bird to clean itself easily after it has burrowed into large jungle fruit when feeding. It is a bird of the lowlands and hill forests up to 2000m. (27)

PNG and Australia are well known for their colourful parrots and a good example is the Dusky Lory *Pseudeos fuscata*. It is a very common bird found only on New Guinea and some nearby islands from the lowlands up to the lower montane forests at 1800m. It usually moves around in flocks and is very noisy when feeding in flowering trees. It is well known in some coffee plantations where it feeds on the flowers of the *Leucaena* grown to shade the coffee. (28)

The White Cockatoo *Cacatua galerita* is well known to most people. Its feathers are used extensively in headdresses and it figures largely in traditional carvings. It is usually seen singly or in small parties of two or three and not in large flocks as is found in Australia. It is a lowland bird keeping well away from human habitations. (29)

Eclectus Parrots *Eclectus roratus* are usually seen singly or in pairs and are easily located by their raucous calls. Although the sexes are differently coloured they look the same when seen against the sky - just silhouettes. The male is bright green as in the illustration, the female is bright red and blue. (30)

This bird, the Red-cheeked Parrot *Geoffroyus geoffroyi* looks very like the African lovebird parrot. It is only 23cm long. It is a lowland and hill-forest bird and moves around in pairs or small parties feeding in the top of the forest canopy. In the evening many birds meet to roost communally. (31)

28

29

30

31

32

34

33

35

Tourists are fascinated when they see the Papuan Frogmouth *Podargus papuensis* as shown here. At rest this bird is difficult to distinguish from a stump of wood or broken branch. It is nocturnal and hunts insects, mice and bats. Its deep mournful monotone call is a common sound in the bush at night. (32)

Very few people see the owlet-nightjars. They are completely nocturnal and the only time you know they are about is if they give their call - a descending series of bubbling notes. This one is Archbold's Owlet-Nightjar *Aegotheles archboldi* and occurs in the upper montane areas above 2000m in open forest and shrublands. (33)

The long-legged Grass Owl *Tyto capensis* is very common in the open highland valleys where it hunts small rodents in the grasslands and village gardens. Its very long legs and dark dorsal plumage distinguish it from the short legs and creamy yellow dorsal plumage of the Barn Owl found in the same general area. (34)

Sooty Owl, *Tyto tenebricosa.* This is one of the commonest owls in Papua New Guinea, particularly in the denser rainforests from the coast up to the lower montane forests of 1600 m. Here, a mother looks after her two half-grown chicks. The adults make a wide range of calls with high-pitched twitterings and one call which has been likened to the scream of a falling bomb. (35)

One of the best-known birds in PNG is Blyth's Hornbill or Kokomo *Rhyticeros plicatus* but it is seen only in the remoter forest areas. The male, as shown here, has the rusty red colouring of the head and neck; the female is black all over except for the white tail and the bluish white throat. (36)

PNG has a wide variety of kingfishers, several of which are endemic to the island of New Guinea such as the strange Shovel-billed Kingfisher *Clytoceyx rex*. This bird digs into the soft soils of the forest floor to find its food of worms, insects and their larvae. It is a lowland bird but is nowhere common. (37)

The Hook-billed Kingfisher *Melidora macrorrhina* is both crepuscular and nocturnal and often travels through the forest in small parties of maybe ten birds and, at times, all giving their trilling call, each bird's call overlapping another to give a continuous symphony of sound. It is found, or heard, in all lowlands and hill forests of PNG up to 1200m. (38)

One of the smallest kingfishers is the Dwarf Kingfisher *Ceyx lepidus* usually found in forest near streams or even perched over the water. Small insects and arthropods are caught in the foliage or at the surface of streams. It occurs commonly in the lowlands and occasionally as high as the lower montane forests at 1200m. There are two other small kingfishers with similar blue and orange colouring so check your field guide. (39)

37

38

39

40

41

42

43

44

Campbell's Fairy Wren *Malurus campbelli* is the last new species of bird to have been discovered in PNG and this is the first time it has been photographed in a natural setting. Some scientists consider it a race of the Broad-billed Wren found in the Sepik area. It was discovered by Robert Campbell in the Mt Bosavi area just a few years ago. A large area of PNG has not been explored, bird-wise, so there is still the possibility of finding new forms. (40)

The male of the Emperor Wren *Malurus cyanocephalus* as shown here is a very striking bird. The female is white underneath and red-brown on the back and with a blue head like the male. It occurs in the lowlands and moves through the undergrowth in small family groups. It feeds on small insects and can be whistled-up very close so good observations can be made. (41)

A group of birds known as cuckoo-shrikes or greybirds is found in all types of habitat from sea level to very high altitudes in the mountains. This one, the White-bellied or Papuan Cuckoo-Shrike *Coracina papuensis*, is one of the most widespread species being found in all coastal areas and in the hill forests as high as 1800m. A feature of most species is the habit that when they perch they cross their wings, scissors-fashion, two or three times. (42)

The Pittas are a strange group of ground birds which hop along the ground instead of walking or running. They are insectivorous. The Black-headed or Hooded Pitta *Pitta sordida* shown here is partly migratory and for the drier season, April to November, it is common in the lowlands and in the hill forests up to 1200m. (43)

The Blue-breasted Pitta *Pitta erythrogaster* spends most of its time on the ground but is often seen perched in trees where it gives its ground-dove-like mournful call. It breeds in February in the middle of the wet season. The nest is a large dome-shaped structure built on a sloping bank and there are three young. (44)

The tiny - 10cm - Flyeaters or gerygones are often recorded by birdwatchers as 'LBJ's' (little brown jobs).They move so fast through the forest and canopy that by the time you have raised your binoculars to look at them they have moved off as a small brown streak. The Large-billed Gerygone *Gerygone magnirostris* shown here builds a pendant nest with a side entrance with a small hood over the entrance, not clearly defined here. (45)

There are three species of jewel-babblers which replace each other altitudinally. This one, the Blue Jewel-babbler *Ptilorrhoa caerulescens* is found in the lowland forests up to 800m. Like the other species it is a ground bird but is very shy and while you seldom see it you can hear its strong ventriloquial whistle frequently in the forest. (46)

A small bird of the high mountain shrublands, the Wattled Ploughbill *Eulacestoma nigripectus* is well-known to bird-banders for the strength of its bite which easily draws blood from a finger. It uses its bill to break into branches to get to insects and their larvae. The use of the pink wattles is not known though it may have something to do with courtship as the female does not have them. (47)

The Canary Flycatcher *Microeca papuana* is a small bird of the mid to upper montane zones of the mountain forests. It feeds in all levels of the forest, favouring the denser areas, by catching flying insects which it sees from a convenient perch. It is not very shy so it is easily seen when birdwatching along forest trails. (48)

Among the group of birds known as whistlers is this small species known as the Dwarf Whistler *Pachycare flavogrisea*. It occurs from the hill forests at 1000m to the lower montane forests of 1500m and sometimes higher. It feeds in the mid-storey of the forest to the canopy and gleans among the leaves for insects. (49)

45

47

46

48

49

50

51

53

52

54

Another of those small brown birds which are difficult to identify is the Mountain Mouse-Warbler *Crateroscelis robusta.* It lives in the high mountain moss-forests above 1700m. There are three mouse-warblers which are difficult to identify unless you realise that they are separated largely by altitude and their calls are very distinctive. (50)

The mention of starlings often brings an adverse reaction to householders. However, we have several native species of starlings which are not pests like the European Starling which does not occur here though there are a couple of records of it from Port Moresby. This one, the Singing Starling *Aplonis cantoroides*, is found in forested areas as well as open grass and shrublands up to 1700m. (51)

The Northern Fantail *Rhipidura rufiventris* is a common forest bird of the lowlands and hill forests up to 1500m. It usually spies an insect from a perch, dashes out to catch it and takes it back to the perch to eat. It also occurs in northern Australia. (52)

The Emperor of Germany Bird of Paradise *Paradisaea gulielmi* is found only in the Huon Peninsula from 500m to 1300m. A special feature of the display of this species is that it hangs upside down and spreads its white flank plumes into a sort of dish. As with the Raggiana B.o.P., the males of this species gather at a particular tree and when a female approaches all the males slip upside down and display to her. (53)

The birds of paradise are our most spectacular and colourful birds. There are thirty-three species (some scientists say thirty-five) in PNG and the Raggiana Bird of Paradise *Paradisaea raggiana* shown here is featured on our national emblem. It is a common bird found in the lowlands from the Fly River in the southwest to the eastern tip of PNG and westwards along the north coast to Lae and Madang. It also occurs in the mountains up to the Wahgi Valley at 1700m. (54)

The metallic-like feathers of the head, neck and breast of the Princess Stephanie Bird of Paradise *Astrapia stephaniae* reflect and refract the sunlight into shimmering variegated colours as the bird turns this way and that. With its long broad black tail it is one of the most magnificent of the birds of paradise. (55)

In the northwestern quarter of PNG lives the Lesser Bird of Paradise *Paradisaea minor*. It occurs from the coastal fringe up to 1600m in the Baiyer Valley. This species also displays in one particular tree in a local area. Up to fifteen males or more will display there early morning and just before dusk. (56)

The Parotias are a group of four birds of paradise, three of which are in PNG, which are characterised by the six wire-like feathers on the head. In courtship these 'wires' are whirled about and look like the blades of a small fan in motion. The Lawes Parotia *Parotia lawesi* is the commonest of the three species here, and like the others it makes a cleared patch on the ground as its display and courtship area. (57)

One of the least-known and less spectacular birds of paradise is this Yellow-breasted Bird of Paradise *Loboparadisaea sericea*. It has a peculiar swollen blue wattle on the top of the bill. Its function is unknown but until recently illustrations of this bird showed this wattle to be two flat or shrunken disks - which happens when the bird dies and is made up into a museum study skin. This is what was used by artists to paint a picture of this bird. In the wild it is rare . (58)

The Brown Sicklebill *Epimachus meyeri* is one of the largest of the birds of paradise. The male has a tail about 66cm long. Its call has been likened to the sound of a machine-gun firing. It is a bird of the high montane forests where it gleans for insects and berries among the leaves and mosses and trunks of pandanus and other moss-forest trees. (59)

56

57

58

59

60

61

63

64

One of the dominant and beautiful calls in the dawn chorus of the lowland rainforests is that of the Hooded Butcherbird *Cracticus cassicus*. It received its name from the habit of some species of killing nestlings of other birds and impaling them on a thorn or broken twig where the bird can feed on them then or at a later time. (60)

The bowerbirds are a very interesting group and the amazing structures built by the males for courting the females are still a wonder to ornithologists. Here we have the Yellow-breasted Bowerbird *Chlamydera lauterbachi* about to enter its bower which is made up of four walls of sticks, the longer side walls fronting this photograph, forming an avenue. (61)

The Ribbontail Bird of Paradise *Astrapia mayeri* was the last of the birds of paradise to be discovered. It was described and named in 1939. Since then it has been found commonly throughout its habitat in the high mountain moss-forests of the Western Highlands, Southern Highlands and Enga provinces. It has the longest tail of any wild bird in the world - about a metre long. (62)

There are several native birds which occasionally use man-made structures as sites for nest-building. Here a Long-billed Honeyeater *Melilestes megarhynchus* has built its mossy nest in the thatch roof of the gatehouse of the Baiyer River Sanctuary. The young bird is fed insects, lizards and worms. As an adult it takes nectar from flowers as well as insects taken from the curled dead leaves of plants where larvae have pupated. (63)

The glistening red plumage of the King Bird of Paradise *Cicinnurus regius* is one of the gems of colour of the lowland forests though difficult to see until it moves. It nests in a tree hollow, unlike other birds of paradise which make cup-shaped nests in the forks of branches. (64)

Among the large number of honeyeaters in PNG the nine species of *Meliphaga* yellow/white-eared birds are most difficult to identify in the field. The Yellow-gaped Honeyeater *Meliphaga flavirictus* shown here is a common species at the Baiyer River Sanctuary in the Western Highlands. In any one area of the coastal fringe of PNG there will be three species to find. (65)

PNG has a counterpart of the Australian Magpie-lark called the Torrent Lark *Grallina bruijni* which, like the former, makes a mud nest. It lives in the hill and montane forests up to 2400m, never far from running water, where it spends most of its time on the ground and in mountain streams, hopping from one rock to another, catching insects. (66)

The pitohuis are endemic to New Guinea and are related to the whistlers, shrike-tits and shrike-thrushes. The Rusty Pitohui *Pitohui ferrugineous* shown here is a common species of the lowlands often associated with other species of birds in feeding flocks which move through the forest disturbing the foliage and revealing insect prey. Its white eye helps to distinguish it from other species. (67)

The Black Butcherbird *Cracticus quoyi* is sparsely distributed throughout the lowlands of the New Guinea island and it is found in north Australia too. It is the largest of the four species found in PNG and its call, though loud, is not as melodious as in the other species. It occurs as high as 1500m where the lowland forests and valleys have intruded into the highlands. (68)

A very common small bird found in the lowland rainforests is the Black Berrypecker *Melanocharis nigra* shown here. Its continuing up-and- down trilling call is to be heard frequently at any hour of the day. It is interesting to watch these birds apparently attempting to do the impossible by attacking a berry much too big for the size of its bill only to see that the berry is so soft that it squashes down to the bird's mouth size. (69)

65

66

67

68

69

Most of the large continents of the world have one species of bustard. These are large ground birds which walk in stately fashion but they are shy birds. In PNG we have the Australian Bustard *Ardeotis australis* but it is only found in the Trans-Fly and Bensbach area. It is a bird of the open plains, savannah woodlands and the edges of swamps and lakes. (70)

The New Guinea Thornbill *Acanthiza murina* is one of our smallest birds. It is our only representative of a large group of birds called thornbills in Australia. It occurs in the highest montane forests and is continuously on the move in small to large flocks. It always appears to be busy finding small insects to eat. It is often difficult to distinguish from the small brown scrubwrens in the same habitat. (71)

One of our most beautiful birds of the high montane forests and shrublands is the Crested Berrypecker *Paramythia montium*. It is a fruit and seed eater sometimes coming down to the ground to feed. When disturbed the bird raises the feathers of its black cap into a crest. It sometimes nests communally. (72)

70

71

72